FUN FACT FILE: HABITATS

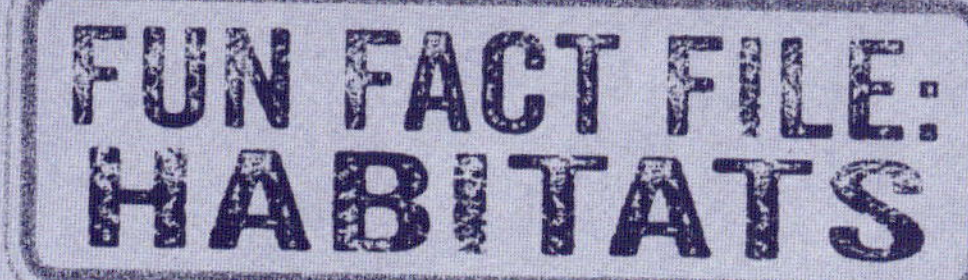

20 FUN FACTS ABOUT MARINE HABITATS

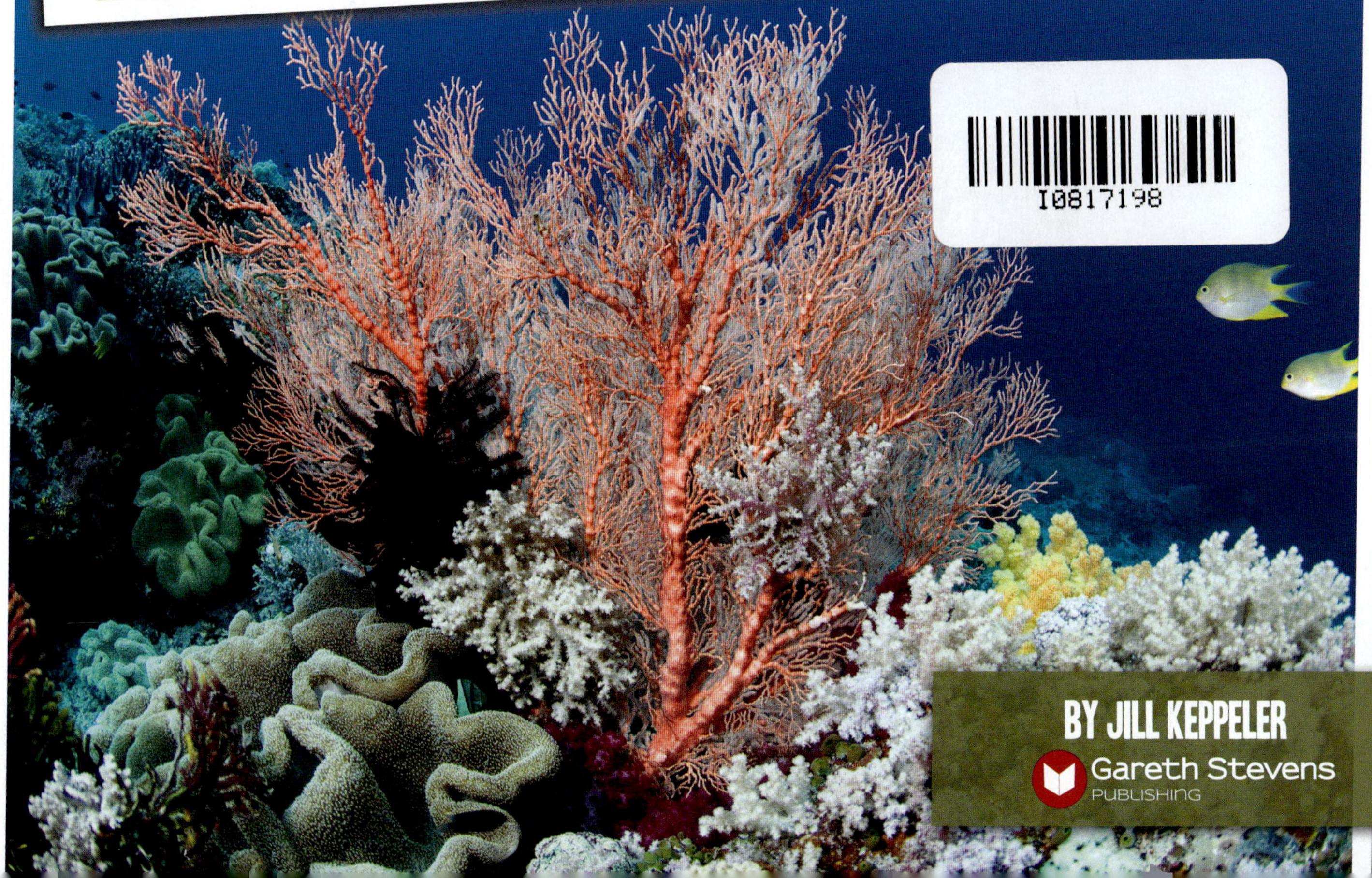

I0817198

BY JILL KEPPELER

Gareth Stevens
PUBLISHING

Please visit our website, www.garethstevens.com. For a free color catalog of all our high-quality books, call toll free 1-800-542-2595 or fax 1-877-542-2596.

Library of Congress Cataloging-in-Publication Data

Names: Keppeler, Jill, author.
Title: 20 fun facts about marine habitats / Jill Keppeler.
Other titles: Twenty fun facts about marine habitats
Description: New York : Gareth Stevens Publishing, [2022] | Series: Fun fact file. Habitats | Includes index.
Identifiers: LCCN 2020036977 | ISBN 9781538264515 (library binding) | ISBN 9781538264492 (paperback) | ISBN 9781538264508 (set) | ISBN 9781538264522 (ebook)
Subjects: LCSH: Marine habitats–Juvenile literature. | Marine ecology–Juvenile literature.
Classification: LCC QH541.5.S3 K455 2022 | DDC 577.6–dc23
LC record available at https://lccn.loc.gov/2020036977

First Edition

Published in 2022 by
Gareth Stevens Publishing
29 East 21st Street
New York, NY 10010

Copyright © 2022 Gareth Stevens Publishing

Designer: Michael Flynn
Editor: Kate Mikoley

Photo credits: Cover, p. 1 (main) Darryl Leniuk/DigitalVision/Getty Images; file folder used throughout David Smart/Shutterstock.com; binder clip used throughout luckyraccoon/Shutterstock.com; wood grain background used throughout ARENA Creative/Shutterstock.com; p. 5 titoOnz/iStock/Getty Images; p. 6 Alexius Sttandio/Shutterstock.com; p. 7 Tammy616/E+/Getty Images; p. 8 picturist/iStock/Getty Images; p. 9 Paul Souders/Stone/Getty Images; p. 10 tonaquatic/iStock/Getty Images; p. 11 eco2drew/iStock/Getty Images; p. 12 by wildestanimal/Moment/Getty Images; p. 13 DEA PICTURE LIBRARY/De Agostini Picture Library/Getty Images; p. 14 Brent Durand/Moment/Getty Images; p. 15 Douglas Klug/Moment/Getty Images; p. 16 ullstein bild/Getty Images; p. 17 courtesy of NASA; p. 18 Barcroft Media/Getty Images; p. 19 VectorMine/iStock/Getty Images; p. 20 best works/Shutterstock.com; p. 21 3dsam79/iStock/Getty Images; p. 22 Felix Martinez/Moment/Getty Images; p. 23 VextorMine/iStock/Getty Images; p. 24 lisnic/Shutterstock.com; p. 25 ullstein bild/Getty Images; p. 26 Chris Strickland/iStock/Getty Images; p. 27 somnuk krobkum/Moment/Getty Images; p. 29 Anna Gorbacheva/iStock/Getty Images.

All rights reserved. No part of this book may be reproduced in any form without permission in writing from the publisher, except by a reviewer.

Printed in the United States of America

Some of the images in this book illustrate individuals who are models. The depictions do not imply actual situations or events.

CPSIA compliance information: Batch #CSGS22: For further information contact Gareth Stevens, New York, New York at 1-800-542-2595.

CONTENTS

Words in the glossary appear in **bold** type the first time they are used in the text.

THE BIGGEST HABITAT

Earth's oceans cover nearly three-quarters of its surface. Marine, or ocean, **environments** are the biggest overall type of **habitat** on the planet! And they're not just widespread. They're sometimes very deep. The oceans have an average depth of about 12,100 feet (3,688.1 m).

Though they all have things in common, there's a lot of **variety** in marine habitats too. They can be warm or almost freezing, **shallow** or deep, close to land or far away from it.

Earth is sometimes called the Blue Planet because of how its oceans look from space.

SO MUCH VARIETY

FUN FACT: 1

SCIENTISTS THINK THAT MORE THAN 1 MILLION KINDS OF PLANTS AND ANIMALS LIVE IN EARTH'S OCEANS.

In fact, there may be as many as 9 million more species, or kinds, of life-forms we haven't discovered yet!

Life in Earth's oceans ranges from tiny plants we can't even see to huge creatures such as great white sharks!

Exploring the deepest parts of the ocean can be dangerous, or unsafe, and it can cost a lot of money!

FUN FACT: 2

HUMANS HAVE EXPLORED ONLY A SMALL PART OF THE OCEANS.

There's a lot out there we don't know about! People have also mapped less than 10 percent of the oceans with modern **technology**. We know more about areas near the coasts.

Continental shelf habitats make up only 7 percent of the ocean.

FUN FACT: 3

MOST MARINE LIFE LIVES IN A SMALL PART OF THE OCEAN.

There are two main types of ocean habitats: coastal and open ocean. Most marine life lives in coastal habitats on the **continental shelf**.

FUN FACT: 4

THE TWO MAIN HABITATS CAN BE FURTHER BROKEN DOWN INTO MANY DIFFERENT TYPES.

Temperature, or how hot or cold something is, is one thing that affects marine habitats. Polar habitats are very cold and have a lot of ice.

Many things affect what animals and plants live in an area. Ocean habitats include coral reefs, kelp forests, seagrass meadows, the open ocean, and more.

VERY BIG AND VERY SMALL

FUN FACT: 5

SOME OF EARTH'S SMALLEST CREATURES LIVE IN THE OCEAN.

You can only see some ocean creatures through a microscope, an instrument used to see very small objects. Zooplankton are tiny swimming or floating creatures that serve as food for many other animals.

There are many kinds of zooplankton. There are also phytoplankton, which are tiny forms of plant life.

FUN FACT: 6

THE LARGEST CREATURE TO EVER LIVE ON EARTH LIVES IN THE OCEAN.

The blue whale is bigger than any other animal—it's even bigger than any dinosaur that once lived on this planet.

GROW A BACKBONE!

FUN FACT: 7

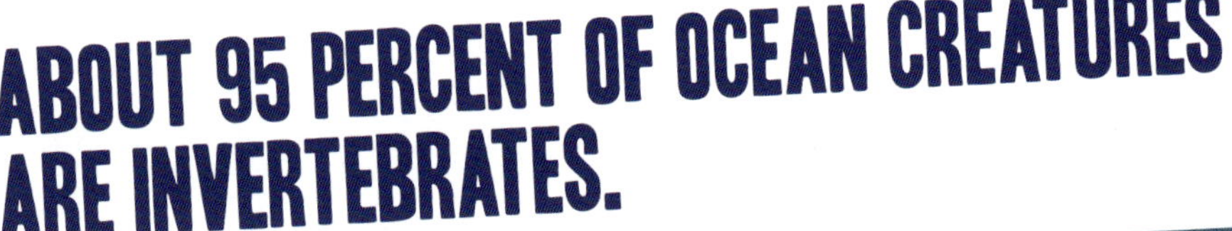

Invertebrates are creatures without a backbone, or a row of bones protecting their spine, a string of **nerves** in their back. Jellyfish and shrimp are invertebrates. So are squids, lobsters, crabs, and clams.

The giant squid is the biggest invertebrate on Earth. It can reach nearly 60 feet (18.3 m) in length.

There are about 30 kinds of bristlemouth fish in the ocean.

FUN FACT: 8

THE MOST COMMON VERTEBRATE IN THE OCEAN—AND ON EARTH—IS A GLOWING FISH!

A vertebrate is an animal with a spine. The bristlemouth is a small fish that glows in the dark. Scientists think there are hundreds of trillions of them in the oceans!

UNDERSEA FORESTS

FUN FACT: 9

THERE ARE HUGE FORESTS OF KELP IN EARTH'S OCEANS.

Just like forests on land, these forests under the ocean surface provide homes and food for many animals. Kelp is a large brown seaweed or **alga**.

Kelp is often found in cold, clear water.

Kelp has no roots. It's stuck to the ocean floor by parts called holdfasts.

FUN FACT: 10

SOME KELP CAN GROW TO BE MORE THAN 250 FEET (76.2 M) LONG.

Giant kelp can grow up to 1.5 feet (0.5 m) a day. Some kinds will stretch upward because they have air-filled bladders, or bubbles, that float.

GRASSY MEADOWS

FUN FACT: 11

THERE ARE MEADOWS UNDER THE OCEAN.

These undersea meadows are made up of seagrasses, water plants that grow almost anywhere. They're the only flowering plants in the ocean, and there are more than 70 species.

This seagrass meadow is located near Indonesia. Seagrass meadows provide homes and food for many animals.

The dark-green-brown marks on this photo taken from space show seagrass beds on the Al Wadj Bank.

FUN FACT: 12

SOME SEAGRASS MEADOWS ARE BIG ENOUGH TO BE SEEN FROM SPACE!

The Al Wadj Bank in Saudi Arabia, on the north coast of the Red Sea, has both huge seagrass meadows and large coral reefs.

OPEN OCEAN

FUN FACT: 13

THE OPEN OCEAN IS ANY AREA AWAY FROM THE COASTAL AREAS. IT CAN BE FAIRLY SHALLOW OR VERY DEEP.

The open ocean is divided into zones—not by where on Earth they are, but by how deep under the water's surface they are.

The biggest ocean species, such as this ocean sunfish, live in the open ocean.

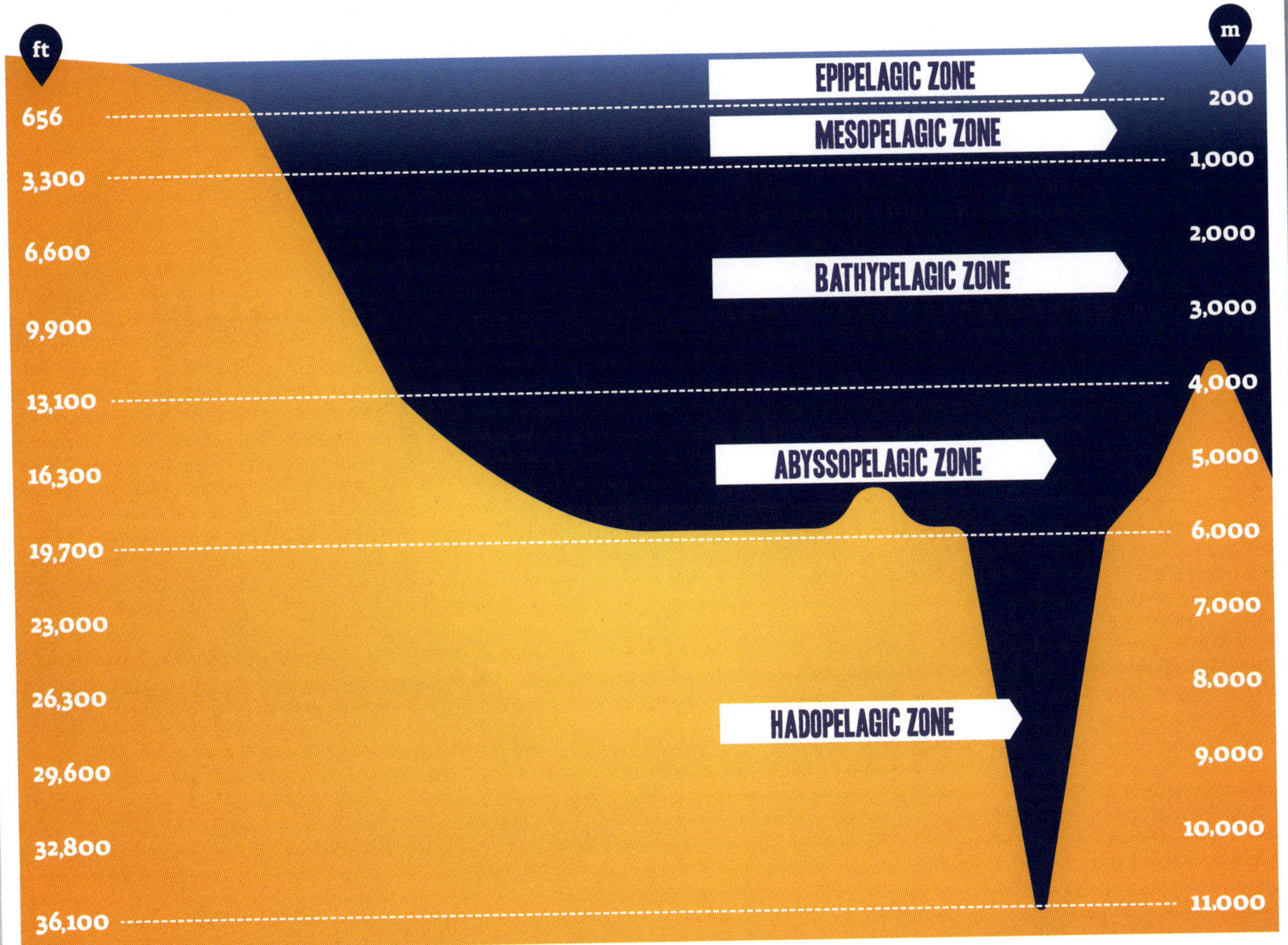

The entire open ocean is known as the pelagic zone. It's further broken up into the zones shown here.

THE DARKEST AND DEEPEST

FUN FACT: 14

PART OF THE OCEAN IS DEEPER THAN MOUNT EVEREST IS TALL!

Mount Everest, Earth's highest mountain, is about 29,000 feet (8,839.2 m) tall. Scientists think the deepest part of the ocean is about 36,000 feet (10,973 m) deep!

The Mariana **Trench**, located in the Pacific Ocean, is the deepest of the deep-sea trenches. There's no light, and the water is nearly freezing.

Goblin sharks can move their jaws, the bones of the face where teeth grow, out of their face to catch food!

FUN FACT: 15

THE MARIANA TRENCH IS HOME TO WEIRD-SOUNDING CREATURES SUCH AS THE DUMBO OCTOPUS, SEADEVIL ANGLERFISH, GOBLIN SHARK, AND ZOMBIE WORM.

Thousands of types of creatures have adapted, or changed, to live in the trench. We probably don't even know about many of them.

CITIES OF THE SEAS

FUN FACT: 16

CORAL REEFS HAVE MORE BIODIVERSITY THAN ANY OTHER MARINE HABITAT!

About 25 percent of the world's marine animals live in coral reefs. However, these reefs only take up less than 1/10th of 1 percent of the ocean floor.

The Great Barrier Reef off Australia is the world's largest coral reef system.

THE WORLD'S CORAL REEFS

ARCTIC OCEAN

NORTH AMERICA

EUROPE

ASIA

PACIFIC OCEAN

AFRICA

ATLANTIC OCEAN

PACIFIC OCEAN

SOUTH AMERICA

INDIAN OCEAN

AUSTRALIA

SOUTHERN OCEAN

SOUTHERN OCEAN

ANTARCTICA

- **THE GREAT BARRIER REEF,** CORAL SEA, NEAR AUSTRALIA
- **RED SEA CORAL REEF,** RED SEA, MIDDLE EAST
- **NEW CALEDONIAN BARRIER REEF,** PACIFIC OCEAN, NEW CALEDONIA
- **MESOAMERICAN BARRIER REEF,** ATLANTIC OCEAN, NEAR MEXICO
- **FLORIDA REEF,** ATLANTIC OCEAN AND GULF OF MEXICO, UNITED STATES

This map shows the locations of some of the largest coral reefs on the planet.

FUN FACT: 17

CORAL REEFS ARE BUILT BY (AND PARTLY MADE UP OF) LIVING CREATURES.

A coral reef starts with tiny animals called polyps. As polyps die, they leave hardened skeletons behind. Other polyps form skeletons over that base.

It takes polyps and their skeletons thousands of years to form the **structure** of a coral reef.

The algae that gives coral its bright colors is called zooxanthellae algae. Millions can live in a square inch (6.5 sq cm) of coral!

FUN FACT: 18

CORAL REEFS' BRIGHT COLORS COME, IN PART, FROM ALGAE.

Coral polyps don't have any color at all, but they draw a kind of algae that lives in them and makes them colorful.

NEAR THE OCEAN

FUN FACT: 19

SOME HABITATS ARE PARTLY OF THE OCEAN AND PARTLY OF SMALLER WATERWAYS OR LAND.

Habitats such as estuaries, or sites where rivers empty into the ocean, are important to wildlife. Marine creatures, land animals, and birds may all make their homes here.

Estuaries have a mix of both fresh water and salt water. Many birds stop at them while **migrating.**

Mangrove trees appear to sit on top of the water, thanks to their tall roots.

FUN FACT: 20

MANGROVE FORESTS ARE HOME TO TREES THAT HAVE ADAPTED TO USE SALT WATER!

These forests, located in warm areas where the ocean meets the land, are home to many creatures. The tree roots help keep coastlines in place.

PROTECTING MARINE HABITATS

The health of Earth's marine habitats is important to the health of the planet as a whole. The oceans help keep Earth from getting too hot or too cool. They're also home to millions of creatures, all with their own part to play in the environment.

However, Earth's oceans are in danger. Pollution and global warming caused by humans hurt marine habitats and their animals. We all need to do our part to keep our oceans safe for all.

Recycle plastic bottles instead of throwing them away. This will help keep plastic out of Earth's marine habitats!

GLOSSARY

alga: a plantlike living thing that is mostly found in water; plural is algae

biodiversity: the different kinds of life in an environment, shown by numbers of different kinds of plants and animals

continental shelf: a shallow, or not deep, area under the ocean's surface around the edge of a continent, or one of Earth's seven great masses of land

environment: the natural world in which a plant or animal lives

habitat: the place or type of place where a plant or animal naturally or normally lives or grows

migrate: to move from one area to another for feeding or having babies

nerve: a part of the body that sends messages to the brain and allows something to feel things

shallow: not deep

structure: something built

technology: using science, engineering, and other industries to invent useful tools or to solve problems. Also, a machine, piece of equipment, or method created by technology

trench: a long, narrow hole in the ocean floor

variety: the state of having many different kinds of things

BOOKS

Davies, Monika. *How Deep in the Ocean?* Mankato, MN: Amicus Illustrated, 2019.

Gardeski, Christina Mia. *All About Oceans.* North Mankato, MN: Capstone Press, 2018.

Hawkins, Emily. *Atlas of Ocean Adventures.* London, England: Wide Eyed Editions, 2019.

WEBSITES

Ocean Habitat
kids.nationalgeographic.com/explore/nature/habitats/ocean/
Learn more about marine habitats and their importance on this National Geographic Kids website.

What Is Happening in the Ocean?
climatekids.nasa.gov/ocean/
On this NASA Climate Kids website, learn more about the importance of the oceans and how they affect us and our planet.

World Biomes: Marine
kids.nceas.ucsb.edu/biomes/marine.html
Explore the plants and animals of the Earth's marine biomes and find more interesting links.

Publisher's note to educators and parents: Our editors have carefully reviewed these websites to ensure that they are suitable for students. Many websites change frequently, however, and we cannot guarantee that a site's future contents will continue to meet our high standards of quality and educational value. Be advised that students should be closely supervised whenever they access the internet.

INDEX